Table Of Contents

Introduction 1

Chapter One: Main Features of IPhone XS and iOS12 1

    OLED screen 1

    A12 Bionic Processor 1

    Neural Engine 1

    Face ID and TrueDepth Camera System 2

    Face ID Security and Privacy 2

    Camera Specs 2

    Video Capabilities 3

    Stereo Recording 3

    512GB Storage Option 4

    Water and Dust Resistance 4

    Fast Charging 4

    Wireless Charging 4

    What to expect from iOS 12 4

Chapter Two: Setting up the iPhone XS 6

    How to turn on your iPhone? 6

    How to turn on a frozen iPhone XS? 6

    Learn the controls on the iPhone 7

    What items do you need to set up your iPhone XS for use? 7

    How to move your data from an Android device to iPhone XS? 8

    Setting up a cellular service on iPhone XS 8

        How to install the Nano-SIM? 8

        How to set up an eSIM? 9

        How to manage your cellular service providers? 9

        How to connect your iPhone XS to the internet? 10

    Management of Apple ID and iCloud settings on your iPhone XS 10

        How to sign in with your Apple ID? 11

        How to change your Apple ID settings? 11

        How to change your iCloud settings? 11

        Features of the iCloud 12

    Setting up the Face ID 12

Chapter Three: Basics 14

Learn basic gestures to interact with iPhone XS 14

Check out the meaning of status icons on iPhone XS 17

Use 3DTouch for previews and shortcuts 19

How to use the iPhone Home screen and open apps? 20

Change your iPhone Settings 20

    Set the date and time of your iPhone XS 21

    Set language and region 21

    How to change the name of your iPhone XS? 21

    Setting up an email, contact and calendar accounts 21

    How to change other settings? 22

    How to change or lock the iPhone XS screen orientation? (For XS Max only) 22

    How to change the wallpaper on your iPhone? 22

    How to increase or reduce the iPhone XS screen brightness and color? 23

    How to Turn True Tone on or off? 23

    How to amplify the iPhone XS screen with Display Zoom? (for XS Max only) 24

    Adjusting volume on your iPhone XS 24

    How to adjust iPhone XS sound and vibrations? 25

Notifications 25

DND 26

How to type and edit text? 28

    Correct spelling 28

    Type with one hand 28

    Typing options 28

    How to choose a text? 29

    Use the keyboard as a trackpad 29

    How to turn off predictive text? 29

    Dictate a text for the iPhone 29

    How to add punctuation to a text? 29

Customize and use Control Center 30

Open and organize Today View 31

Set the Screen time allowances, and limits 31

Siri 33

How to backup iPhone using iCloud or iTunes? 34

    Setting up iCloud 34

    Setting up iPhone with iTunes 34

Charge the battery 34

    How to find out battery charge level remaining in the status bar 35

    Turn on the Low Power Function 35

    Some additional info about your battery applying 36

    Check your battery condition 37

Chapter Four: How to use various apps 37

    App basics 37

        How to switch between Apps? 38

        How to move and organize your app? 38

        Ensure your favorite apps are readily accessible on iPhone 38

        Customize You Control Center to include your favorite apps 39

        Use 3DTouch for quick actions from the Home screen 39

    App Store 39

    Books 40

    Camera 41

    FaceTime 46

        How to set up FaceTime? 46

        How to make calls with FaceTime? 47

        How to manage FaceTime calls? 47

    Files 48

    Find My Friends 49

    Find My iPhone location 50

    iTunes 51

    Safari 53

Chapter Five: Tricks and Tips on how to use the iPhone XS 53

Introduction

The latest line of smartphones do not showcase as many remarkable changes or game-changing attributes as past years. However, when linked to the iOS 12 rollout of Sept. 17, 2018, the devices offer plenty to users looking to upgrade.

Whether it is the Xs or the Xs Max, chances are, you're keen to turn it up and start exploring the new features. **Note**: Functionality wise the Xs and Xs Max are identical, so although we focus here on the Xs model, all is applicable for the Xs Max version.

Chapter One: Main Features of IPhone Xs and iOS12

OLED screen

The iPhone Xs offers the same 5.8-inch OLED display as the iPhone X, but the Xs's "Super Retina" display has a higher pixel count and 60 percent greater dynamic range than its predecessor. The iPhone Xs Max's 6.5 inch screen sets a new record with 3.3 million pixels.

A12 Bionic Processor

The iPhone Xs comes with a 7-nanometer A12 Bionic chip. This chip is faster than the A11 chip found in the previous version of the iPhone X. The A12 also enables a longer battery lifespan.

Neural Engine

The Neural Engine can carry out five trillion operations within a second. This allows you to perform more functions in real time and facilitates the performance of such learning functions as photo search, Face ID, etc.

Face ID and TrueDepth Camera System

Face ID was introduced in 2017 with the iPhone X. Face ID performs a similar role to the Touch ID with the only difference being the use of the facial scan. The A12 Bionic chip increases the speed at which the Face ID works to a matter of seconds. Face ID works through sensors and the built-in camera at the rear of the iPhone Xs. This system is referred to as the TrueDepth Camera system. The camera makes use of a Dot Projector that projects more than 30,000 invisible dots on your face. These dots are read by an infrared camera.

Face ID Security and Privacy

Face ID makes use of a high quality 3D facial scan. It is impossible to trick this facial scanner as it focusses on the eyes to work. This works when the "Attention Aware" option is turned on. If your eyes are closed, it won't work. You can also turn it off if you choose. However, using it provides additional security.

The attention aware features let the iPhone Xs know when you are looking at it. You can disable Face ID quickly without a thief knowing - all you need to do is tap the Side button and the volume button at the same time.

The Face ID locks after trying to recognize your face twice. You will need to insert the passcode to reopen it.

The interesting thing about Face ID is that even Apple can't access your Face ID data. In other words, nobody can. The Face ID data is encrypted in an Enclave on your iPhone.

Fooling Face ID is a 1 in 1,000,000 chance. Although, the possibility of identical twins unlocking the Face ID cannot be ruled out, the conclusion is that an average person cannot unlock your iPhone Xs.

Camera Specs

Apple's 'Smart HDR' (similar to Google's 'HDR+') combines the best of multiple photos taken at different exposures into a single image. This significantly improves dynamic range, while improvements to its

Portrait Mode can adjust background blurring of an image after it is taken.

In addition to new six-piece lenses, there's one other major upgrade in the iPhone Xs:

- Primary rear camera - 12MP, f/1.8 aperture, 1.4µm pixel size, Optical Image stabilisation (OIS), Quad-LED True Tone flash, Portrait Lighting

- Secondary telephoto lens - 12MP, f/2.4 aperture, 1.0µm pixel size, OIS, 2x optical zoom

- Front 'TrueDepth' camera - 7MP, f/2.2 aperture. Perfect for taking selfies. With this front camera, you can capture detailed images. The Depth Control feature allows you to adjust the amount of blur after taking the image or when composing the image in real time.

Video Capabilities

If you are a photo freak, then you have just got yourself an advanced camera. The iPhone Xs has the ability to do the following;

- To record 4K video at 24,30 or 60 frames within a second.
- To record a 1080p HD video at 30 or 60 frames within a second
- To record 720p HD video at 30 frames per second.
- It has a 3x digital zoom when you're capturing a video.
- The camera supports a 1080p Slo-mo video.

Stereo Recording

In addition to the more powerful camera and overall quality improvements thanks to Smart HDR, videos will now capture stereo sound via one of four microphones built into the iPhone Xs for a improved sound experience.

512GB Storage Option

Experience double the storage with the 512GB model for those who want to pay a few hundred dollars extra. The 256GB model is still available for those with less storage demand.

Water and Dust Resistance

A unique feature of the iPhone Xs is IP68 water and dust resistance. This means you can submerge your iPhone in up to three metres of water.

Avoid intentional immersion of your iPhone Xs in water. Please note - Apple's warranty does not cover damage caused by international water damage to the iOS device.

Fast Charging

The iPhone Xs comes with a 5W charger. If you use your iPhone often, you will get a 50 percent charge within 30 minutes. You need a USB-C power adapter that produces 18 watts. You can purchase a charger for your iPhone Xs from a licenced Apple store.

Wireless Charging

This an advanced way of charging an iPhone without actually plugging it in the usual USB-C power adapter. You can use a Qi wireless to charge your iPhone Xs. Several reviews have shown that wired charging works faster than wireless charging.

What to expect from iOS 12

The iPhone Xs runs on the iOS 12. This application has revolutionised what an iPhone is.

The main focus of iOS 12 is to increase performance. Some new features include Memoji, Screen Time, Group FaceTime and so much more. Developers can also make use of the iOS 12.

Check out some amazing features on iOS 12.

Dual SIM – You can make use of a Dual SIM on the iOS 12 system. You can either use a nano-SIM and an eSIM (unavailable in some countries).

Group FaceTime – This feature supports chatting with up to 32 participants at the same time (specific regions only).

Camera – Supports depth adjustment after taking a Portrait.

Apple Music – You can easily browse, discover, and listen to your any music of your choice. You can also visit new artist pages and view the top 100 songs. Check out what your friends are listening to on Friends Mix.

Performance – iOS are produced to work faster and is more responsive.

Messages – Creating a personalized Memoji is made easier. Express yourself with added effects, filters, and stickers. The new photo iMessage app helps you get photo suggestions.

FaceTime – This affords you a lively conversation with the use of an Animoji or Memoji.

Screen Time – This feature provides an insight into how your friends make use of the iPhone.

Notifications – Notifications are grouped. This makes it easier to see all the information.

Do not Disturb – When you activate features such as Do Not Disturb Bedtime will dim all displays and silence all notifications during bedtime.

Measure – Just point your iPhone camera at a real-world object to get the dimensions.

Photo – It has a smarter features than the previous iOS.

Siri – You can add shortcuts for regular activities. You can ask Siri to perform the activities. As soon as Siri learns your routines, it will offer suggestions of what you can do.

iOS 12 can carries out updates automatically.

Chapter Two: Setting up the iPhone Xs

This chapter will take you through from the moment you first see "Hello" on your device screen to the last step. In other words, you will have a perfect understanding of how to set up your iPhone Xs yourself.

How to turn on your iPhone?

This may seem easy for regular iPhone users, but it wouldn't be fair to a new iPhone user to skip this part. It's very simple. Just press and hold the Side button until the iPhone turns on. If it doesn't, then you may need to charge the battery. In very rare cases will the iPhone not turn on after plugging it in to charge. Such a situation is referred to as a frozen screen.

How to turn on a frozen iPhone Xs?

If your iPhone Xs has a frozen screen or doesn't react to touch, here is what you should do;

1. You may need to force a restart. This doesn't wipe the iPhone's memory, rather, its aim is to turn on the phone. To force a restart, do the following;

- Quickly press and release the volume up button.
- Quickly press and release the volume down button.
- Lastly, press and hold the Side button until the Apple logo shows on the screen and a recovery screen mode appears.
- Press the option to update. The iTunes will attempt a re-installation of iOS. This process will not erase your iPhone memory.

2. Plug the iPhone and let the battery charge for 1 hour. Ensure that the charger is plugged firmly into your socket and the iPhone during this process.

7

Learn the controls on the iPhone

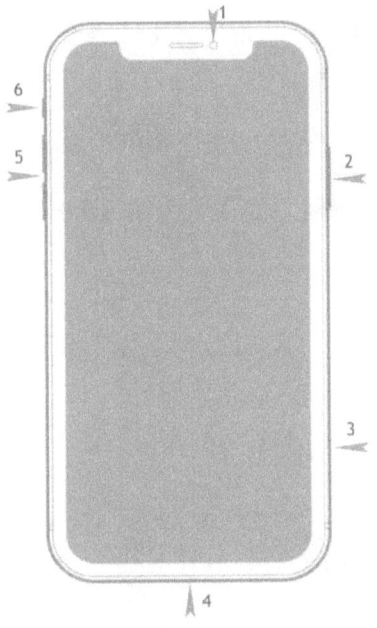 

1. Front-facing cameras

2. Side button

3. SIM tray

4. Lightning connector

5. Volume buttons

6. Ring/Silent switch

7. Rear cameras

8. Flash

What items do you need to set up your iPhone Xs for use?

You need the following items to be available;

1. An internet connection via a cellular data service or a Wi-Fi network.

2. An Apple ID and password. You can create an Apple ID during set up.

3. A credit or debit card account details. You need this if you are adding a card to your Apple Pay account.

4. The backup data on your previous iPhone or Android device.

How to move your data from an Android device to iPhone Xs?

Are you one of those smartphone users who feel that it's not possible to move your data from your Android device to an iPhone Xs? Simply, follow these easy steps to carry out this task;

For your Android device

1. Turn on your cellular data or Wi-Fi network.

2. Download the Move to iOS app on your Android device.

3. Follow the instructions.

For your iPhone Xs

1. Follow set-up assistance.

2. On the screen, tap moves data from Android.

Setting up a cellular service on iPhone Xs

To set up a cellular connection, you need a SIM card. You can contact your service provider to help you set up a data plan. The iPhone Xs supports a Dual SIM. It makes use of a nano-SIM and an eSIM (this may not be available in some countries).

How to install the Nano-SIM?

1. Push the SIM tray to eject.

2. Gently remove the tray from the iPhone.

3. Insert the nano-SIM into the tray.

4. Push the tray into the iPhone gradually.

How to set up an eSIM?

1. Tap Settings > press "add cellular plan".

2. Put the iPhone position that will allow the QR code provided by your service provider to appear. You can also insert the QR code manually.

3. Press the "Add Cellular Plan".

4. You can also follow the onscreen instructions to set it up how you want it.

Alternatively, you can download your service provider app on the Apple store. It will help you set up the eSIM faster and easier. Do you know that you can store than one eSIM on your iPhone? However, you can only use at a time.

How to manage your cellular service providers?

This part will help you set up your cellular plan the way it suits you. Your settings will determine how iPhone Xs makes use of each cellular service provider.

Go to settings > tap Cellular and follow these instructions;

1. Press the Cellular and pick a service provider.

2. You can change settings such as Cellular Plan Label, Calls, SIM pin code.

If you want to register your phone number for iMessage and Face Time, go to settings, tap iMessage and Face Time and follow the instructions. You can't register more than one number for iMessage and Face Time.

Note the following if you're using Dual SIM;

1. Any incoming call when you're on the other network will automatically go to voicemail. You may not receive a missed call notification. However, this will only happen when the voicemail service is activated.

2. The calling waiting only works if the incoming call is on the same network.

3. Switching conversations from one network to another is not possible. Then, if you must, you may incur charges for your service provider.

How to connect your iPhone XS to the internet?

An iPhone without an internet connection is worthless. You need an internet connection for virtually everything on this device. The exception is taking pictures and making videos. Let's take a look at how to connect your device to the internet.

Wi-Fi connection

1. Settings > Tap Wi-Fi > Select Network > Insert password

2. Tap the network to insert the password.

As soon as this sign 📶 appears, your Wi-Fi is turned on.

Cellular Connection

The cellular data network becomes active as soon the Wi-Fi connection is switched off.

1. Check your SIM setting to ensure that is activated.

2. Settings > Cellular.

3. Tap the Cellular button to ensure that it turns on.

Management of Apple ID and iCloud settings on your iPhone Xs

Your Apple ID is like the "password" you need to do almost anything on your iPhone Xs. If you don't have an Apple ID, you can't store content on iCloud, purchase contents from iTunes Store, App Store, etc.

iCloud helps you store your photos, videos, documents, music and so on securely. It also helps you share pictures, locations, music and lots more to your loved ones. You can use the iCloud to find your iPhone when it's lost, keep reading, you will find out about this shortly.

iCloud also gives you a free email account and 5 GB to store your mail, pictures, videos, etc. You can also upgrade your iCloud Storage space from your iPhone.

Please note that you may not have access to all the features on iCloud if you reside in some regions and the iCloud features may vary from one region to another.

How to sign in with your Apple ID?

Assuming you didn't sign in during setup, follow these instructions;

1. Settings > Click Sign in > Insert your Apple ID and password

2. You don't have to worry if you forgot your password, just go to Recover Apple ID website (https://support.apple.com/en-us/HT201487).

How to change your Apple ID settings?

Go to iPhone Settings (enter your name) and do any of the following;

1. Update your contact information.

2. Change password.

3. Manage family sharing.

How to change your iCloud settings?

Go to settings > (your name) > iCloud, you can do any of the following;

1. Check your iCloud memory.

2. Upgrade your iCloud memory

3. Turn on the features you need.

Features of the iCloud

The iCloud is a unique feature that you can only find on Apple's iPhone. Here are some features of iCloud that will interest you;

1. You can use the iCloud to update your messages, mail, contacts, calendars, photos, videos, music, app, books, contents, passwords and credit cards.

2. Share images and videos with your friends.

3. You can locate iOS, Apple watch, or Mac devices that belong to you or a friend.

4. Back up your contents.

5. Share your location with your friends and family.

Please note that enabling iCloud for apps like music, photos, and contacts will stop you from using iTunes to synchronize them to your PC.

Setting up the Face ID

When you launch your iPhone Xs for the first time, you will be asked to set up the Face ID. You may choose to set it up at a later time. Please have it in mind that Face ID can only memorize one face. It doesn't matter when you choose to set up the Face ID, and the process is the same. The entire process is easy and fast.

Go to Settings > Face ID & Passcode, enter Password. Then the image below will appear. Just tap on "Get Started" to start the process.

Position your face for the front camera to capture you. Immediately, the sensor recognizes that there is a face ready to scan, you will hear an interesting background that will let you know that your face is about to be scanned.

On the screen this will display "Move your head slowly to complete the circle". Just do a slow neck roll while you hold your iPhone Xs. This will help the iPhone Xs to map all the angles and corners on your face. You will also hear interesting animation sounds as you move your face in a circle.

As you keep moving your face, a green circle will cover your face.

You will have to go through the process twice. This is to ensure that all angles of your face are scanned properly.

When the process is completed, tap the Done tab.

Face ID Options

After setting up the Face ID, you also need to choose when it should function. As a default setting, you must look at the iPhone XS directly to unlock it. The fact that you have to look at the iPhone directly means that somebody cannot snatch your iPhone and hold it up to your face when you're not paying attention.

If you have a disability or wear glasses, just turn off the "Require Attention for Face ID". You will find this in the Settings > Face ID & the Passcode section.

If Face ID identifies that you are looking at the screen, it will dim the screen or play the message alerts or whatever notifications at a low volume. However, you can disable this feature off when you tap the "Attention Aware Features".

You can also choose the section you want to use the Face ID to function. For instance, you can turn it off for Apple Pay, third-party apps, App store, etc.

You can set up an alternative appearance. For instance, when you are wearing a hat or a sunglasses. The process is the same as mentioned earlier. Just tap "Set up Alternative Appearance".

Although the Face ID is remarkably accurate, there are a few instances that it may struggle to identify your face. If you notice such, you can do the following;

Move a little bit closer to your iPhone.

Take off your sunglasses. The Face ID will not recognize your face if you're putting on a sunglasses that block infrared.

Step away from a very bright light.

As you keep using the Face ID, it will also identify your face even if you change your hair style or mustache. If the Face ID is not sure it's you, it will ask for your passcode to unlock.

Chapter Three: Basics

Let's start with some iPhone Xs favorites.

Learn basic gestures to interact with iPhone Xs

The iPhone Xs has some simple gestures that will help you control the device and use all apps effectively.

	Tap - touch your finger lightly on the screen
	Press - press the screen firmly with one finger only
	Swipe - move your finger across the screen swiftly

	Scroll - move your finger across the screen without removing it
	Zoom - place two of your fingers close to each other on the screen of the iPhone Xs. If you want to zoom in spread them apart, then to zoom out move them towards each other. Another way to zoom is to tap the screen. You can zoom in double tapping and also do the same to zoom out
	Go Home - move your finger from the bottom of the screen to go back to the Home screen
	Quickly access controls - move your fingers from the top right hand side of the screen to access the Control Center. If you want to add or remove items, Go to Settings > Control Center > Customize Control

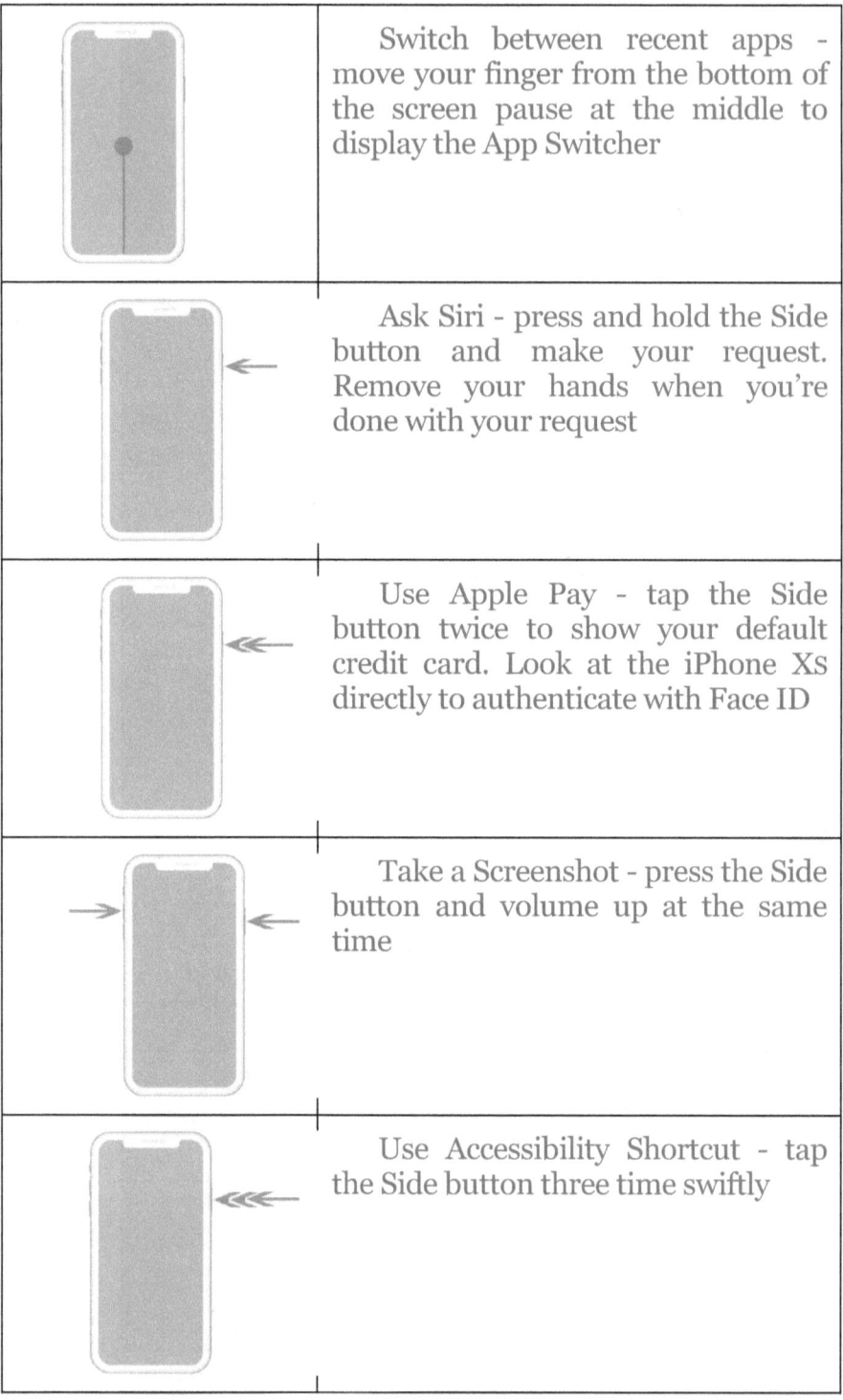

	Switch between recent apps - move your finger from the bottom of the screen pause at the middle to display the App Switcher
	Ask Siri - press and hold the Side button and make your request. Remove your hands when you're done with your request
	Use Apple Pay - tap the Side button twice to show your default credit card. Look at the iPhone Xs directly to authenticate with Face ID
	Take a Screenshot - press the Side button and volume up at the same time
	Use Accessibility Shortcut - tap the Side button three time swiftly

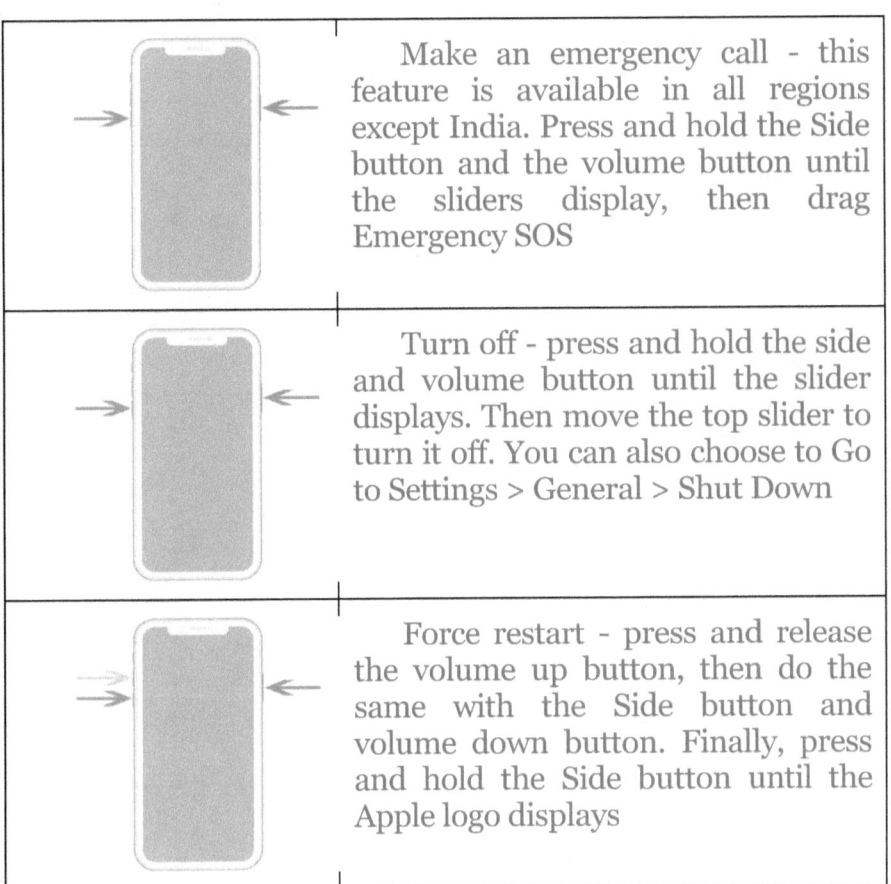

		Make an emergency call - this feature is available in all regions except India. Press and hold the Side button and the volume button until the sliders display, then drag Emergency SOS
		Turn off - press and hold the side and volume button until the slider displays. Then move the top slider to turn it off. You can also choose to Go to Settings > General > Shut Down
		Force restart - press and release the volume up button, then do the same with the Side button and volume down button. Finally, press and hold the Side button until the Apple logo displays

Check out the meaning of status icons on iPhone Xs

The icon that display in the status bar helps you know the app is turned on or off.

📶	Cell signal – This signal informs you how strong or weak the strength of your cellular service is. If there is no network signal, "No Service" will display in place of the cellular signal
📶📶	Dual cell signals – The first cellular signal bar shows the strength of the first line on your iPhone while the lower signal bar is on the second line
✈	Airplane mode – If you see this sign, it shows

	that the Airplane mode is activated. This means that you can't make or receive calls or perform any wireless functions
LTE	This sign indicates that your LTE network is available. This feature is not available all the regions
4G	This signal shows that 4G UMTS network is available. This is not available in some regions
3G	This signal indicates that you can access 3G UMTS (GSM) network
E	EDGE – If this feature appears, then it means that EDGE network is available
GPRS	The iPhone Xs can connect to the GPRS
Wi-Fi	If the Wi-Fi call is activated, you can make calls with your iPhone Xs
🛜	This signal means that your Wi-Fi is connected
⦾	This sign shows that your Personal Hotspot is connected to another device for use
↻	If this signal appears, it means that your iPhone Xs is syncing with iTunes
☼	It shows that network activities are turned on. Some third-party apps may also use this signal to show that it is active
(→	It means that the Call Forwarding is turned on
VPN	It shows that you are connected to a VPN network

☎	This signal shows that software RTT/TTY is turned on
🔒	It means that your iPhone Xs is locked
🌙	Means that Do not disturb is turned on
🔒↻	Shows that the iPhone screen is locked in portrait orientation
➤	As soon as you see this, it means that Location Service is being used
⏰	It shows that an alarm is turned on
🎧	It indicates that a Bluetooth headphone is turned and paired to your iPhone Xs
🔋	It displays the battery level of a Bluetooth device paired with your iPhone Xs
🔋	It lets you know the battery level or charging status of the iPhone Xs. If yellow appears, it means that Low Power Mode is on

Use 3DTouch for previews and shortcuts

To show the quick actions menu on the screen press any icon on the Home screen.

To set the brightness of the screen you should open the Control Center, press or touch and hold the flashlight icon and drag the slider.

When you see a notification on the Lock screen, press it to respond.

Typing the text you can make your keyboard work like a trackpad.

To make a line darker during drawing you can press the appropriate icon in Notes.

Using the Photos, press an icon a bit stronger and you will be able to see the picture on the full screen. It's enough to press the picture, go up and you will be able to share or copy it if needed.

The Mail application. If you received a message press the icon in your mailbox to discover its content. Move it up to see a list of actions.

How to use the iPhone Home screen and open apps?

The Home screen of any phone including the iPhone Xs displays all the apps on the device. Follow the instructions below;

1. Swipe up from the bottom edge of the iPhone Xs screen.

2. Move left or right to go through apps on your Home screen pages.

3. If you want to open an app, tap it.

4. To go back to the Home screen, swipe up from the bottom edge.

Change your iPhone Settings

The Settings of the iPhone Xs helps you to configure and customize the device. You can choose your preferred language, sounds, notification, etc.

You can do the following through the Settings;

Set the date and time of your iPhone XS

The default time of the iPhone is set based on your location. However, you can change it if they are incorrect.

Go to Settings > General > Date and Time.

You can also do any of the following;

• Set Automatically: This option allows the iPhone XS to set the time and date as provided by your network. However, in some regions, some networks don't support automatically setting the time and date of your iPhone.

• 24-Hour Time: This feature is not available in all regions.

Set language and region

To carry out this function, do the following;

1. Go to Settings > General > Language and Region.

You can also set the following;

• The language of the iPhone.

• Region.

• The Calendar format.

• The temperature unit.

2. If you want to add a keyboard for another language, go to Settings > General > Keyboard > Keyboards.

How to change the name of your iPhone XS?

Tap Settings > General > About > Name.

Touch ⊗, insert the name of your choice and press Done.

Setting up an email, contact and calendar accounts

The iPhone XS works with Microsoft Exchange and other familiar Internet – based mail.

Tap to Settings > Passwords & Accounts > Add Account.

Adding an email account requires that you tap any service provider. E.g. Google, Yahoo, etc.

How to change other settings?

Tap Settings > swipe downwards > insert a term e.g. alert, password, music etc.

How to change or lock the iPhone XS screen orientation? (For XS Max only)

Most apps on the iPhone XS have a different layout when the device is in landscape orientation. Such apps are Mail, Message, and Photos.

Please *note* that these different layouts are not accessible when you Zoom in.

Unlocking or locking the screen orientation

If you don't want the screen orientation to change when the iPhone rotates, lock the screen orientation. Simply launch the Control Center and touch 🔒.

How to change the wallpaper on your iPhone?

Select the image you want as your wallpaper for the Lock screen or Home screen. You can use a dynamic or still image.

Tap Settings > Wallpaper > Select a New Wallpaper.

You can also do any of the following;

• Select any preset image from a group at the top of the screen.

• Choose any of your pictures from your album or photo.

If you want to reposition the photo you selected, double click the screen to zoom in or out. Move the photo with your finger to reposition it.

Press Set, then select from any of the following where the wallpaper should display.

• Set Lock screen

- Set Home screen
- Set as both

Do you know that you can also change the viewing angle of your screen? Simply press Perspective after you have selected a new wallpaper.

Simply tap Settings > Wallpaper > press the image for the Lock screen or Home screen > touch Perspective.

How to increase or reduce the iPhone XS screen brightness and color?

When the iPhone XS brightness is reduced, it helps your battery last longer. You can adjust the brightness manually.

Tap Control Center and drag ☼

Tap Settings > Display and Brightness > slider down or up to reduce or increase.

To adjust the screen brightness automatically

Tap Settings > General > Accessibilty

Press Display Accomodation > switch on Auto Brightness.

How to Turn True Tone on or off?

This feature when turned on lets the iPhone to automatically adapt to the color and intensity of your environment.

Launch the Control Center > tap ☼ > press ☼ > to switch on or turn off True Tone.

Tap Settings > Display and Brightness > tap True Tone on or off.

Program the Night Shift to switch off and turn off automatically

The Night Shift feature enables your iPhone to display a warmer color at night. This will help you see clearly when it's dark.

- Tap Settings > Display and Brightness > Night Shift.
- Switch on Scheduled.

• If you want to change the color balance of the Night Shift, move the Color Temperature to the warmer or cooler part of the spectrum.

• Press From and choose Sunset to Sunrise or Custom Schedule.

If you selected Custom Schedule, press options to program the times that the Night Shift should go on or off.

If you also selected Sunset to Sunrise, your iPhone XS will make use of the information from your clock to know when it's night time in your region.

Take *note*: The option for Night Shift from Sunset to Sunrise will not appear if your Location is turned off.

How to amplify the iPhone XS screen with Display Zoom? (for XS Max only)

To magnify what's displayed on your screen, do the following;

• Tap Settings > Display and Brightness

• Press View

• Select Zoom and press Set.

Switch on and use Reachability

• Tap Settings > General > Accessibility > Reachability.

• To get the top of the screen, Swipe down to the bottom edge of the screen.

If you want to reset the screen, press the top of the screen.

Adjusting volume on your iPhone XS

You can adjust the volume of the iPhone via the Volume button. You can find it on the side of the device. Siri is also there to help. Just ask Siri to either reduce or increase the volume it will be done immediately.

Adjusting volume on your iPhone XS in Control Center

Another way to adjust your iPhone XS volume is through the Control center.

Launch Control Center > drag 🔊).

When the device is in ring mode, it plays every sound. However, in silent mode, the iPhone doesn't produce any sound or alert. The iPhone will vibrate even it's in silent mode.

Note: clock alarms and other apps like Music will play sounds even when the iPhone XS is on silent mode.

How to adjust iPhone XS sound and vibrations?

You can adjust or change the sounds iPhone plays when you get a call, email, reminder, and other notifications.

Setting sound and vibration options

• Tap Settings > Sounds and Haptics.

• Drag the slider either above Ringers and Alert to adjust the volume for all sounds.

• To adjust the vibration pattern and tones, press the type of sound you want; e.g. ringtone or text tone.

Note: If the iPhone doesn't make any sound when you receive incoming calls, go to the Control Center and check if Do Not Disturb is turned on. If you see 🌙 icon, tap it to put it off.

Notifications

Notifications will keep you posted on what's new. You can also customize your notifications to view what's important to you.

You will receive all the notifications as soon as they arrive. Assuming, you don't open it, all notifications will be saved. To view all the notifications, swipe down from top of the screen. To close the Notification, swipe up.

When there are multiple notifications, they are grouped by app. That way, it is easier to view and manage. If the notification is grouped, it will display as small stacks and the most current notification will appear on top.

Do any of the following;

• You can expand the group notifications to view them individually.

• Touch the notification to view it.

• If you want to respond to the notification, swipe to the right and open it. To see notification content you should press hard on it and now you are able to answer it.

You can either dismiss or clear your notifications. There are ways to manage your notifications;

• When you are using another app and receive a notification, swipe down to view and swipe up to dismiss.

• To clear your notifications, swipe to the left-hand side > press clear or clear all.

• To send notifications to the Notification Center, swipe to the left.

• Whenever you want to turn off notification for an app, swipe to the left > touch Manage > press to turn off.

To clear all your notifications in the Notification Center, touch .

When an app has not been used for a long time, Siri often suggests that that the app is turned off.

Tap Settings > Notifications > Siri Suggestions > turn off.

DND

Set up "Do Not Disturb"

This function quickly silence your iPhone if you don't want to receive calls or notifications. You can ask Siri to turn on Do Not Disturb.

Go to Control Center > touch .

To turn it off also press .

How to turn on "Do Not Disturb" while Driving?

Turning on "Do Not Disturb while Driving" helps you to drive safely without any distraction by silencing messages and notifications. Siri will read the replies aloud so that you don't have to look at the iPhone. You will only receive calls when the car Bluetooth system is turned on.

• Tap Settings > Do Not Disturb.

• Move down > press Activate.

You can do any of the following;

• Turn on "Do Not Disturb while driving" to detect when you're driving automatically. This means that "Do Not Disturb while driving" will be turned on as soon as it detects that you are driving.

• Turn it on when Connected to Car Bluetooth.

• You can also turn it on manually.

Note: If CarPlay is activated, "Do Not Disturb while driving" will not work.

Sending an auto-reply that you are driving

When activate "Do Not Disturb while driving", it will auto-reply any message sent to you.

• Tap Settings > Do Not Disturb > Auto-Reply To.

You can do any of the following;

• No One: Turn off auto-reply.

• Recent: The iPhone will auto-reply to anyone who you sent a message previously.

• Favorite: Your device will send an auto-reply to your favorite group on your iPhone.

Assuming you receive a message with the heading "Urgent", the remaining texts will go to the reminder drive.

If you want to enjoy a bedtime without any interference, set "Do Not Disturb during Bedtime". It will reduce the display, reduce calls, all through the night until the morning.

How to type and edit text?

There are so many apps that require the use of text. Some of them are Contacts, Messages, Notes, Mail, etc.

You can do the following while inserting text;

- Press Shift and touch the letter to use Uppercase letters.
- Tap the Shift key twice swiftly to turn on Caps Lock.
- Press **123** to enter numbers or Symbols such as #+=.
- To Undo the last edit, shake the device and press Undo.
- To use emoji, touch 😃 or 🌐.
- To use accented letters or other characters, press and hold the key.

Correct spelling

If your text is underlined in red, it means that the spelling is wrong.

Touch the underlined word to view the correct spelling or suggestions.

Touch the suggestion to replace the underlined word.

Type with one hand

To use one hand to type your text, move the keys closer to your thumb.

Press and hold 😃 or 🌐.

Slide and select one of the keyboard layouts.

Typing options

If you don't want a spelling check to be on, you can turn it off.

Press and hold 😃 or 🌐 > move to Keyboard Settings.

Typing options that you can turn on or off are Auto-Capitalization, Auto-Correction, etc.

How to choose a text?

Touch and Hold the text > allow the magnifying glass to show > drag to place an insertion point.

Use the keyboard as a trackpad
Press the keyboard until the color of the keyboard becomes light gray.

Slide the insertion point around the keyboard.

You can carry out any of the following as you type a text;

You can accept a suggested word or emoji.

Reject a suggestion by touching the original text.

How to turn off predictive text?

Press and hold 😃 or 🌐.

Go to the Keyboard Setting and switch off Predictive.

Inasmuch as you turn off the Prediction, the iPhone Xs will still provide suggestions of misspelled text. You can either accept or decline. If the rejection is consistent, iPhone Xs will stop suggesting.

Dictate a text for the iPhone

If you don't want to type a text, you can detect it. However, this feature may not be available in all regions.

Tap Settings > General > Keyboard.>Turn on or off Enable Dictation.

Press 🎤 in the onscreen keyboard.

Read out the text.

When you are through, press ⌨.

How to add punctuation to a text?

While you are dictating the text, you also have to read out the punctuation. For instance, "Hello Alice comma the check is in the mail exclamation mark full stop". Some of the punctuation and formatting commands;

Quote ... end quote

New line

Cap – to capitalize the next word.

Smiley – to insert :-)

Frowny – to insert :-(

Customize and use Control Center

When you go to Control Center, it will give you access to Do Not Disturb, flashlight, etc.

- Open Control Center > tap the top-left group of controls > press 📶 > AirDrop options.
- Open Control Center > tap 📷 to take a selfie.
- Open Control Center > press 📶 to reconnect.
- If you want to see the name of the Wi-Fi network, tap 📶
- To switch off Wi-Fi tap Settings > Wi-Fi. (Switch on Wi-Fi in the Control Center and press 📶.)
- To connect to Bluetooth open Control Center > press ✴.

- Tap Settings > Control Center > turn off Access Within Apps.
- If you choose to customize controls, tap Settings > Control Center > Customize Controls. Tap ≡ next to a control move it to a new position.

Open and organize Today View

Today View helps you get an information from favorite apps. You can view today headlines, calendar, etc.

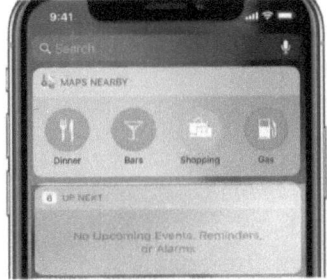

- Tap Today View > press Edit below the screen.
- If you want to remove widgets, press ⊕ or ⊖.
- If you want to adjust the pattern of the widgets in Today View, tap ≡ > move it to a new position. Tap Settings > Face ID & Passcode > Insert your passcode >Switch on Today View.

Set the Screen time allowances, and limits

You can set how allowances and screen time are used. Screen Time will provide a summary of how long you spend on each app on your iPhone. It will also provide information about the apps that send you notifications regularly. This information will assist you to set limits for some apps and websites.

Open your Screen Time summary

As soon as you set up the Screen Time, it will prepare a summary of how you use your iPhone Xs. It will give you an insight into how you use the apps. If you want to see your Screen Time summary, tap Settings > Screen Time > insert the name of your device. The summary will contain the following;

• The amount of time you spend on each app.

• The breakdown of how you use the app by the time each day.

• The time you spent on each app and the app that you used above the time limit.

• A summary of all the notifications you get.

• Which app you use frequently.

Setup Screen Time for yourself

This will afford you the opportunity to set allowances and limits for each app. Here is how to set it up;

Tap Settings > Screen Time.

Press Turn On Screen Time > touch Continue > press This is My iPhone.

- You can carry out any of the following:
- You can set up the start and end times by turning on Downtime.
- Press App Limits > Add Limit > choose one or more categories.
- Press Always Allowed > press ⊕ or ⊖ close to the app to either include it or remove it.

Get a report of your device use

To view your Screen Time report, follow these steps:

Tap Settings > Screen Time.

Tap your device name or All Devices near the top of the screen.

You'll see your summary for today.

Tap Last 7 Days to view a summary of how you used the device over the week.

Siri

Siri is a voice assistant. Ask Siri to translate a phrase, calculate, set a timer, find a location, about the weather, and more. Siri can make calls, send mails, read message etc.

Note: To use Siri, iPhone must be connected to the Internet.

For set up Siri go to Settings > Siri & Search > turn on Listen for "Hey Siri" for summon Siri with voice/Press Side button for Siri for summon Siri by press and hold Side button.

Go to Settings > Siri & Search > My Information > Tap your name.

Try ask Siri something like:

Hey Siri, followed by

- Turn up the volume
- Turn on Do Not Disturb
- Press and hold Side button and ask
- How's the weather today?
- What's 15 times 5?
- Set up a meeting with Alice at 10
- What's on my calendar for Monday?
- What time is it?
- What time is it in Bangkok?
- Wake me up tomorrow at 7 a.m.
- What's my sister's work address? (if you have added your sister's contact details)
- Hey Siri, what can you do?

How to backup iPhone using iCloud or iTunes?

The essence of iCloud is to help you backup your information. It backs up your iPhone daily when it is connected a Wi-Fi to power.

Setting up iCloud

Tap Settings > [insert your name] > iCloud > iCloud Backup.

Switch on iCloud Backup.

Press Back Up Now for manual backup.

To delete a backup, select backup and press Delete Backup.

Setting up iPhone with iTunes

Use a USB to connect your iPhone Xs your PC.

Open iTunes on your PC.

Tap the iPhone tab, tap Summary in the sidebar.

Choose "Encrypt iPhone backup" to encrypt the backup

When this sign 🔒 appears it means that the backup is encrypted.

You can tap any of the following;

Back Up Now: This will prompt a manual backup.

This computer: Automatic backups start whenever the iPhone is connected to your computer.

Whenever you want to view or delete iTunes backups, select iTunes > Preferences > tap Devices.

Charge the battery

To charge your iPhone battery, you have to perform one of the following actions:

- With the help of the Lightning to USB Cable and USB power adapter, which are included, connect your iPhone to a power point.

- Switch on your computer. Join your iPhone to a USB port on it. Synchronize your iPhone with iTunes. In this case your computer has to be turned on. If your computer is switched off, the battery will discharge instead of charging. Pay attention to the battery icon ⚡ to assure your iPhone is charging at the moment.
- It's preferable to use an iPhone face up on a Qi-certified charger.
- When you connect your iPhone to a power outlet or put it on a Qi-certified charger, an iCloud backup or wireless iTunes syncing can be turned on.

Note: In case your keyboard does not have a high-power USB port it's impossible to charge your iPhone.

You can follow the battery level or charging status with the help of battery icon located in the upper-right corner of the screen. In case you synchronize or use iPhone at the same time with the charging, it may take much time to charge the battery.

You can see when the battery is nearly discharged with the help of special display image. It will take about 10 minutes to charge the battery and you can use iPhone again. If iPhone is almost entirely discharged the display may be blank for about 2 minutes. Then the image of low-battery will appear.

Warning: It's forbidden to use the Lightning connector to charge iPhone near liquid.

How to find out battery charge level remaining in the status bar

With your finger swipe from the top-right corner down the screen.

Turn on the Low Power Function

This function can noticeably make the life of the battery charge longer. It's advisable to use Low Power Mode in case your iPhone battery is almost discharged or you don't have any eans to recharge shortly.

Press Settings > Battery >Turn on Low Power Function.

Low Power Function restricts background activity and adjusts performance for the most important actions such as incoming or outgoing calls, email, and messages; available Internet; and others.

Note: *I*f your iPhone switches to Low Power Function automatically, it goes back to usual power function after it reaches to 80% battery charge. It may take more time to fulfill some tasks in case your iPhone uses Low Power Function.

Some additional info about your battery applying

Go to Settings > Battery.

Info about your battery charging and performance will be shown for the period from last 24 hours to the last 10 days.

- *Insights and suggestions:* It will help you to find out what conditions or schemes of applying make iPhone to take energy. You may also find the tips for decreasing energy consumption - press to go to the corresponding setting.
- *Last Charge Level:* Informs about the level of battery and when it was disconnected.
- *Battery Level schedule (in Last 24 Hours):* Informs you about the battery level, time when it was charging, and when iPhone used the Low Power Mode or the battery was extremely low.
- *Battery Usage schedule (in 10 Days):* Informs you about the percentage of battery used during each 10 days.
- *Activity graph:* Informs you about operation during interested periods, depending the screen was switched on or off.
- *Screen On and Screen Off:* Informs you about general activity for the chosen time interval, considering when the screen was switched on and when it was switched off. The Last 10 Days view informs you about the mid per day.
- *Battery Usage by App:* Informs you about how much battery each application used during the chosen time interval.

- *Activity by App:* Informs you about how much time each application used during the chosen time interval.
- *Remember:* To find out battery information for chosen period, press that time period in the graph. To cancel it, press outside the graph.

Check your battery condition

Settings > Battery >Press Battery Condition.

On the display of your iPhone you can find information about your battery's output, the most productivity, and when your battery have to be maintained.

Eventually, output and productivity of all rechargeable batteries, go down. The IPhone's is no exception. If your battery's condition declines an Apple Authorized Service Provider can change the battery to make the productivity and output of high quality again.

Chapter Four: How to use various apps

What makes the iPhone Xs fun, is the numerous and advanced apps that are available. Carefully read the instructions before firing up iPhone Xs apps.

If you have never used an iPhone before, you will be blown away by the numerous apps it offers. You also have the privilege of buying more at the Apple Store. However, the aim of this chapter is to give you a guide of how to use the various inbuilt apps such as Face ID, Smart HDR, camera, Animoji and Memoji on the iPhone Xs.

App basics

The part will help you with the fundamental features that you need. Unlike other iPhones, iPhone Xs doesn't have a Home Screen button. You will learn how to move around the various apps.

How to switch between Apps?

Use the App Switcher to switch to another app. If you switch, you will also pick up from where you stopped. Follow these steps below;

Tap the App Switcher to see all your open apps.

Swipe up from the bottom edge and stop in the middle of the screen.

If you want to check out open apps, swipe to the right.

How to move and organize your app?

This feature helps you organize your app so that you easily locate it. Use the following method to move and rearrange your apps.

Touch and hold an app on the screen until it shakes.

Then drag it to any location of your choice.

You can also create a folder to group all your apps easily.

Touch and hold any app until it shakes.

Drag the app onto another application.

Then drag other apps to the folder.

If you want to name or rename, press the name field and insert the name of your choice.

To delete a folder, drag all the apps in the folder and the folder will be deleted immediately.

Lastly, to reset the Home screen to the previous setting, go to Settings > General > Reset. Then press "Reset Home Screen Layout".

Ensure your favorite apps are readily accessible on iPhone

Keep your favorite apps convenient in Control Center or Today View. In Control Center, shortcuts provide quick access to apps like Notes or Voice Memos. In Today View, widgets offer timely information at a glance. Perform common app functions from the Home screen.

Customize You Control Center to include your favorite apps

Add shortcuts to apps such as Calculator, Notes, Voice Memos, Wallet, and more.

Go to Settings > Control Center > Customize Controls.

Tap the ⊕ icon next to each app you want to add.

Use 3DTouch for quick actions from the Home screen

Press the app icons on the Home screen to open the quick action menus, for example:

- Camera > Take Selfie.
- Maps > Send My Location.
- Notes > New Note.

App Store

You can purchase apps from the App Store, however, you need an Apple ID to sign into the app store.

To find an app, tap "Today" , at the top right tap ⓘ and enter your Apple ID.

You can enter the name of the app on the search tab.

If you want to buy an app, tap the price tag. This sign ☁ means that you have already bought this app and you can download it again without paying.

Tap ⬤ and then press Share App or Gift App.

How to download, view and set restrictions for app purchases?

Tap Today, then your profile picture and tap "Purchased" to view the app you bought.

If the Family Sharing is turned on, press My purchases or select a family member to see the apps they purchased.

To download an app, search for the app you want to download and tap ☁.

To set up app restriction, Go to Settings > Screen Time > Content and Privacy Restrictions.

Books

If you love reading books, then you will find lots of them to read. However, they are not free, you have to buy most of them.

How to find and purchase a book?

You can tap the "Search tab" to find a particular book.

Read a preview of the book or you can include it to your "Want to Read" collection.

Tap "Buy" to purchase a book.

How to read a book on the Books App?

Press Reading Now or Library button to read or see the book you are reading.

To bookmark a page, tap 🔖. To view all your bookmarked pages, tap ≡.

How to listen to a book with Audiobook?

Press the audiobook cover to view the book you want to listen to. Use these buttons to control the audiobook.

☾ - Tap this icon to set the sleep timer.

≡ - Tap to move to another chapter.

How to create a collection, sort, and remove books?

To create a personalized library, tap "Collections" and tap "New Collection".

You can name the collection and tap "Done" when you're through.

If you want to add a book to your collection, tap •••. Choose the collection and add the book.

To sort for a book, tap Library > Sort > Recent, title or author.

To remove a book tap Library > Edit > tap the item you want to remove, then tap 🗑.

Camera

The camera has several modes like Pano, Square, and Portrait, Live Photos, etc. You can either tap the Home Screen or swipe to the left hand side to open the camera from the Lock screen.

Take a Picture

To use the flash, tap ⚡ or you choose auto.

To zoom in or out, pinch the screen.

To select a mode, swipe the screen to the left or right to choose a Photo, Pano, or Square mode.

If you want to take a selfie, tap 🔄.

You can also tap a volume button to take a shot or tap the 📷.

- How to take a Panorama picture?

Select the Pano mode.

Press the Shutter button.

Pan gradually in the path of the arrow.

Tap the Shutter button to finish.

If you want to pan in the vertical direction, rotate the iPhone Xs.

- How to use a picture with a filter?

Select Photo, Square or Portrait mode and press 🎨.

Underneath the viewer, swipe the filters to the right or left.

- How to take a photo in Portrait mode?

Select the Portrait mode.

Use the yellow portrait box to frame your subject.

Then press the Shutter button.

You can remove the Portrait mode effect on the iPhone Xs.

- How to adjust the Portrait Lighting?

This feature of the iPhone Xs gives you the opportunity to edit your photo and add "studio" touches.

Select the Portrait mode and frame your subject.

Choose from any of the following effects;

- Natural light: You can use this effect to sharpen the photo against a blurred background.
- Studio light: Used to brightly light the photo and make it have a clean look.
- Contour light: Adds shadows with highlights and lowlights.
- Stage light: Adds a spotlight on the photo.
- Stage light mono: it has a similar stage light.

Note: You use Stage Light and Stage Light Mono when you front-facing TrueDepth camera.

- How to adjust depth control on Portrait mode?

Select the Portrait mode and frame your subject.

Press ⓕ. You can find it at the top right hand of the screen.

Move the slider to the right to enhance the blur effect. If you want to reduce the blur effect, move the slider to the left.

Press the Shutter Button to take a picture.

- How to take a Burst photo?

This feature will help you take a several photos in high-speed. That way, you can select the best and delete others. The rear and front cameras support this feature.

Select Photo or Square mode.

Press and hold the Shutter tab to take several photos in high speed. To stop, remove your finger.

Press the Burst thumbnail to choose the photos you like.

Press the circle in the bottom right corner of the photo you want to keep and press "Done".

If you want to delete all the Burst photos, press the thumbnail and the delete icon 🗑 .

- How to take a Live Photo?

This feature is spectacular. A Live Photo implies capturing the event that takes place before and after taking the photo, plus the audio.

Select the Photo mode.

Press ◉ to turn on Live photos.

Press the Shutter Button to take a photo.

To edit Live Photos, use the Photo app. Live Photos in your album are marked "Live" at the side.

Note: All images are in HDR (high dynamic range). This feature helps you get take high quality pictures. You can choose to turn this on or off by going to Settings > Camera > Smart HDR. If you turn off HDR, the images will be normal.

- How to record a video?

Some people say that iPhone Xs is superb for recording video. You're about to find out. To record a video do the following;

Select the video mode and tap 📷 to record a video.

You can take a picture while recording a video by pressing the white Shutter button.

To zoom in or out, tweak the screen.

You can either tap 📷 to stop or a Volume Button.

The default setting of the frame rate per second is 30 fps. Go to Settings > Camera >Record Video. If the frame rate is faster, it will increase the resolution of the video. The size of the video file will increase too.

- How to record a video in slow-motion?

This camera mode has the tendency to make feel like you're shooting a movie. Here is how to use it;

Select the Slo-mo mode.

Press the Record button. You can also take a picture while the video is recording.

Press the video thumbnail and tap Edit to set a portion of the video to record in slow motion.

Go to Settings > Camera > Record Slo-mo to change the settings.

- How to take a time lapse video?

Select the Time-lapse mode.

Position the iPhone where you want to capture the event over the particular duration.

Press the record button to begin.

To turn on Auto Low light FPS, go to Settings > camera > record video.

HDR - Go to Settings > Camera > turn on/off Smart HDR.

For save both the HDR and non-HDR versions go to Settings > Camera > turn on Keep Normal Photo.

- How to view, share, and print photos?

All photos and videos are saved in your Photo app. If the iCloud Photos is turned on, every picture is uploaded and accessible from any device that is connected to the iCloud Photos.

To view Photos

Press the thumbnail image at the bottom of the left side.

You swipe to the right or left to view the images that you took.

Press the screen to display or hide the controls.

To share and print photos

Tap ⬆️.

Choose an option either Mail, Messages or Print.

FaceTime

The FaceTime app, allows you to make video and audio calls to your loved ones. It doesn't matter if they use an iOS or Mac device. The front camera of the iPhone Xs allows you to talk to the other person face to face. You can use the rear camera to show your environment to the other person.

FaceTime features may not be available to some users residing in certain regions.

How to set up FaceTime?

Go to Settings > FaceTime and put it on.

To snap Live Photos while you making FaceTime calls, switch on the FaceTime Live Photos.

Insert your phone number, Apple ID, email address to access FaceTime.

Since the iPhone Xs has a Dual SIM, you can change the phone you are using on FaceTime. Go to Settings > Messages > iMessage and FaceTime Line to select the network. You are not allowed to use more than one network at any given time.

How to make calls with FaceTime?

If you have your iPhone Xs connected to a Wi-Fi, a cellular data connection or an Apple ID, you can make and receive FaceTime calls. To use FaceTime via your cellular connection, go to Settings > Cellular > switch on FaceTime calls.

Press ✛ on FaceTime.

Insert the name or phone number you wish to call and press 🎥 to make a video call. If you want to make an audio call, press 📞.

To make a group call on FaceTime, tap ✛ to add more persons by inserting their name, phone number or Apple ID. Assuming no one picks your call, you can press "Leave a Message" to drop a message or cancel the call.

You can set up a FaceTime call from a message conversation. Press 👤, touch the FaceTime tab.

You can either choose to accept a FaceTime call, decline, send a message or tap "Remind Me"

How to manage FaceTime calls?

There are other functions that you can carry out while using FaceTime. You can use other apps while you're making a FaceTime call or even block calls. You can also take a FaceTime Live Photo while you are on a call with a friend.

To take a FaceTime Live Photo, touch ⭕. However, you must ensure that the FaceTime Live Photo is turned on. To do this, go to Settings > FaceTime.

If you don't want a particular caller to call you, simply block their FaceTime calls.

Go to Settings > FaceTime > Blocked

Press "Add New" to add a particular contact to your blocked calls.

Press the contact you want to block.

To unblock a contact, tap unblock.

Files

This app allows you to open and view documents, images, etc., saved on iCloud drive. You can also view documents, images, etc., that are saved in storage providers like Box and Dropbox.

If you want to view a file or folder, just tap it to open it. Changing how files are sorted will help you know where to find your files. Press "Sorted by" to choose whether it will be sorted by name, date, size, etc.

To find a file or folder, do the following;

Touch search and insert the name of the file.

Start the search.

Open the file.

To add a file to a cloud storage device, download an iCloud app from the Apple Store and follow the instructions.

To switch the iCloud on, Go to Settings > iCloud > then select on.

Invite others to use the file on iCloud

You can invite a friend to view a file on iCloud by sending them the link. You can also allow a friend who has the link to the file to edit it. All you have to do is;

Tap the file.

Press Share and touch .

Select the method for sharing the link.

If you want to restrict a person from editing the document, tap "View only".

Find My Friends

This app is a good way to find your friends and family who use iPad, iPad, iPhone, Apple Watch Series 3. You can share your location with them and it will appear on their map. You can also use this app to get notifications when your friends leave or arrive a location.

How to set location sharing?

Tap Settings > [your name] > then switch on Share My Location.

Choose the device you want to share your location with your friends.

Share Location with friends

Launch the Find My Friends app and do this;

Insert the name of the loved one you want to share your location with.

Choose the friend who appears in AirDrop and choose how long you want share the location.

Your friends or loved one will get a notification informing them that you want to share your location with them.

How to set notifications when your arrive at a location?

Choose a friend and press Notify Me.

You can choose if you want to receive notifications when your friend or loved one leaves or arrives a location.

You can ask Siri to find friends who are nearby.

Find My iPhone location

This feature or app ● helps you a locate your iPhone Xs if it gets stolen. The Find My iPhone app has a feature known as "Activation Lock". This feature ensures that nobody can make use of your iPhone Xs, without your permission. However, before this app can work effectively, the Find My Device app must be turned on. The iPhone must have internet access.

How to activate Find My iPhone?

Tap Settings >[your name] > iCloud > Find My iPhone

Switch on Find My iPhone.

Turn on the Send Last Location. This will enable the iPhone Xs to send it last before the battery runs out.

How to find your iPhone Xs?

Open Find My iPhone on another iOS device or via a PC. On your computer, go to Find My iPhone app on iCloud.

Log in your Apple ID. Choose the device you want to find.

Press Action then touch any of the following options;

Play sound

Lost mode

Erase your missing device

Each of these functions will cause the missing iPhone to play a sound even if it's in silent mode, lock your iPhone with a passcode or erase the entire iPhone memory and set it back to default settings. Either of these modes will make the iPhone useless to the thief.

iTunes

The iTunes Store app helps you add music, movies, Tv Shows to your iPhone Xs. If you reside in some regions, renting movies is permitted.

To have access to these files, you must have an internet connection and an Apple ID. Using iTunes is not permitted in some regions.

- How to find music, movies, and so on?

Press either, Music, Movies, or TV shows.

Insert the name of what you are looking for in the search tab.

Touch the result to get more information.

- How to share music, movies, TV shows, etc.?

Press ↥ to pick a sharing option.

To add to Wish list, press ↥ and touch Add to Wish List.

If you like a song playing you, just ask Siri, "what song is playing?" Siri will provide the name of the song, artist, and other relevant information.

- How to buy and download content?

Press price, and touch buy.

If this sign ⌁ appears instead of the price, it means you have bought the item before.

You can check out the progress of the download by taping more.

- How to rent movies?

Each movie lasts for 30 days. You can play it so many times within 48 hours after watching the movie. As soon as the rental period expires, the movie will be deleted.

Press the movie rental price.

Press Rent.

You can either stream the movie or download it.

You can continue to stream or downloading on another device, log in your iTunes ID and Apple ID.

- How to redeem or give an iTunes gift?

Press Music.

Scroll down a little.

Press Redeem or Send Gift.

- Managing your iTunes Store

You can manage how purchases are made from your iTunes. This is very important with regards to purchases that are made within a restricted age in Family Sharing. Family sharing allows you to review and approve items bought by other family members.

Press More and touch Purchased.

If the Family Sharing is activated, pick a family member to view the items they have bought. (this is possible only when the family member decides to share their purchases with you).

Press Music, Movies, or TV shows.

Look for the item you want to download. Press ☁.

To view your entire purchase, tap Purchase history.

Changing iTunes Store Settings

Tap Settings > [your name] >iTunes and App Store to change your settings.

Safari

The Safari web browser allows you to browse the internet. You can also save bookmarks and your browsing history. The Safari browser has similar features like other browsers.

Press the top edge of the screen twice to get back to the top of the page.

To view the page better, turn the iPhone to the landscape orientation.

Press ↻ to refresh the page.

To have a full view of the website press ⬆ and tap Request Desktop Site.

Press ⬆ to share a link.

There are *other brilliant apps such as* Calculator, Calendar, Clock, Compass, Contacts, Health, Maps, Messages, Music, TV, etc.

Chapter Five: Tricks and Tips on how to use the iPhone Xs

The iPhone Xs has an edge to edge design and other amazing features, some which you already know.

Although the previous chapters have explained how to navigate your new iPhone, these tricks and tips will make it even easier. In no time you will master your iPhone Xs.

Tap the screen or press Side button or raise iPhone to wake up.

Swipe the screen upward to view Home screen

If you are looking for a home button on iPhone Xs, you won't find any. Simply swipe up from bottom home bar to take you to the Home screen.

How to access App Switcher quickly?

If you want to access the App Switcher, just swipe up from the left side of the iPhone at a 45 degree angle.

Open Notification center

Just swipe down from the notch area to view all your notifications.

How to open Control Center?

All you have to do is swipe down from the right side of the iPhone XS to access the Control Center.

Easy access to Apple Pay

To use the Apple Pay app, press the Side button twice. The Side button referred to is on the right hand side of the iPhone. As soon as you press the Side button, put the iPhone XS for Face ID scan.

How to switch apps?

To do this, swipe to the right to view the previous application you were using. If you want to go back to the app you were originally using, swipe to the left.

How to take a screen shot?

Since the iPhone XS doesn't have a Home button, just press the Side button and the Volume Down button to take a screenshot

You can view screenshot in the Screenshots album in the Photos app, or in the All Photos album if iCloud Photos (Settings > Photos).

Use Siri

If you want to use Siri, just press and hold the Side button.

How to reboot the iPhone XS

To carry out this function, press the Side button and hold any of the Volume buttons (either up or down).

How to do a hard reset?

This is just like a reboot. No need to worry, it won't delete or wipe any info. Just press the Volume Up and Down button, then hold the Side button till the Apple logo appears.

How to manage two pane landscape view?

If you are using any application, just turn your iPhone sideways to activate a two-pane view.

Get a Home button

The swipe gestures can be quite difficult if you're not used to it. However, if you want a Home button, use the AssistiveTouch feature. Simply go to Settings > General > Accessibility > AssistiveTouch.

Safari

Press the top edge of the screen twice to get back to the top of the page.

You use Airprint to print emails, images, etc. The iPhone and the Airprint must be on the same Wi-Fi network.

Press ⤺ or ⬆ > then Print.

Scanning QR quickly

Apple added a QR code scanner in the camera app. The iOS 12 makes it easier to scan a QR code easier and quickly.

Go to Settings > Control Center > Customize Controls > Scan QR code.

Automatic OS updates

This will help you keep your device up to date without manual aid.

Go to **Settings > General > Software Updates > Automatic Updates.**

From the Lock screen

Open Camera: Swipe left.

See Today View: Swipe right.